ORIENTAL WEEVILS SPECIES

A Comprehensive Guide to Biology, Ecology, and Pest Management

Lawrence Munger

Table of contents

Introduction

What Are Oriental Weevils?

The Oriental weevils are a subset of the vast beetle family Curculionidae, renowned for their unique anatomy and diverse ecological roles. These small insects, often overlooked due to their size, have a significant presence in both natural ecosystems and human activities. Characterized by their elongated snouts or rostrums, Oriental weevils represent a remarkable example of evolutionary adaptation.

Primarily found in tropical and subtropical regions, these weevils exhibit a wide range of behaviors, diets, and ecological interactions. While some are beneficial to the environment, others have earned a reputation as destructive pests. Understanding Oriental weevils requires delving into their dual nature, uncovering the reasons they have fascinated scientists and challenged agriculturalists alike.

Overview of Weevils

Weevils are among the most diverse groups of beetles, with over 97,000 species identified globally. They thrive in various environments, from forests to farmlands, and exhibit an impressive array of forms and behaviors. All weevils share a distinctive feature: their rostrum. This elongated snout not only gives them a unique appearance but also serves as a specialized tool for feeding and egg-laying.

Weevils can be divided into several subfamilies, each adapted to specific ecological niches. Oriental weevils are particularly prominent in regions with warm climates, where they interact with a wide range of plants. Their activities, whether beneficial or harmful, make them integral to the ecosystems they inhabit.

- ## Unique Features of Oriental Weevil Species

Oriental weevils possess several characteristics that set them apart from other beetles:

1. Rostrum Functionality: The rostrum is not just for show; it is a versatile tool used to bore into plant tissues for feeding or laying eeggs.

2. Compact Size: Most species are small, often less than a centimeter in length, enabling them to access hard-to-reach parts of plants.

3. Camouflage: Their coloration and patterns often mimic their surroundings, providing excellent natural defense against predators.

4.Resilience: Oriental weevils are remarkably adaptable, capable of thriving in diverse habitats, from lush rainforests to cultivated fields.

These unique traits have allowed Oriental weevils to flourish in many regions, contributing to their ecological success and, in some cases, their notoriety as pests.

Why Study Oriental Weevils?

Importance in Ecology
Oriental weevils play crucial roles in their ecosystems:

- **Plant Population Control**: By feeding on seeds, leaves, and stems, they naturally regulate plant populations.
- **Nutrient Recycling:** Their consumption of decaying plant matter aids in nutrient cycling, enriching the soil.
- **Food Web Contributions:** As prey for birds, reptiles, and other predators, they occupy a vital niche in the food chain.

Studying these insects provides valuable insights into ecosystem dynamics, helping us understand how various species coexist and interact.

- **Economic and Agricultural Significance**

The impact of Oriental weevils on agriculture is a double-edged sword:

- **Pests of Economic Importance**: Some species, like the banana weevil (Cosmopolites sordidus), are notorious for their destructive habits, causing significant losses in staple crops.
- **Stored Product Infestation**: Weevils often invade stored grains and seeds, reducing their quality and value.
- **Potential for Biocontrol**: On the positive side, certain species may aid in controlling invasive plants or serve as natural regulators of other pest populations.

Understanding the biology, behavior, and interactions of Oriental weevils is essential for devising effective management strategies. Whether it involves protecting crops or preserving biodiversity, this knowledge bridges the gap between ecological needs and agricultural demands.

Part I: Taxonomy and Classification

Chapter 1: The Taxonomic World of Weevils

Introduction to Weevil Taxonomy

Taxonomy, the science of classifying organisms, is fundamental to understanding the diversity and relationships among species. In the case of weevils, their immense variety presents a fascinating challenge to taxonomists. With over 97,000 species recorded globally, the family Curculionidae is one of the largest in the animal kingdom, showcasing remarkable adaptability and specialization.

Oriental weevils belong to this diverse family, with species distributed primarily in tropical and subtropical regions. Their classification relies on detailed observations of their anatomy, behavior, and genetic makeup.

The Family Curculionidae

The family Curculionidae encompasses the true weevils, easily recognizable by their elongated snouts and clubbed antennae. This family is divided into several subfamilies, tribes, and genera, reflecting the incredible diversity within the group.

- **Key Characteristics of Curculionidae:**

- **Elongated Rostrum**: A hallmark feature, used for feeding and oviposition (egg-laying).
- **Clubbed Antennae**: Typically elbowed and positioned on the rostrum for sensory input.
- **Herbivorous Diets**: Most weevils feed on plants, making them significant in both ecological and agricultural contexts.

- **Taxonomic Hierarchy of Oriental Weevils**

The classification of Oriental weevils follows a systematic hierarchy:

Kingdom	Animalia
Phylum:	Arthropoda
Class	Insecta

Order	Coleoptera (Beetles)
Family	Curculionidae
Subfamily	Varied, including Dryophthorinae (common for Oriental species)
Tribes and Genera	Numerous, depending on specific characteristics.

For example, the banana weevil (Cosmopolites sordidus), a well-known Oriental species, belongs to the subfamily Dryophthorinae and the tribe Rhynchophorini.

Challenges in Weevil Taxonomy

1.Morphological Similarity: Many weevil species appear nearly identical, complicating their classification.

2.Cryptic Species: Some Oriental weevils exhibit subtle differences that are only detectable through genetic analysis.

3.Rapid Evolution: The evolutionary adaptability of weevils often leads to the emergence of new species.

To address these challenges, modern taxonomists employ advanced tools like molecular phylogenetics, which uses DNA sequences to determine evolutionary relationships.

Significance of Taxonomy in Understanding Oriental Weevils

Proper classification of Oriental weevils provides a framework for:

- Identifying pest species and their potential threats.
- Understanding ecological roles and interactions.
- Developing targeted pest management strategies.

By delving into their taxonomy, we gain insights into the evolutionary history and ecological significance of these fascinating insects.

Chapter 4: Notable Oriental Weevil Species

Introduction

Oriental weevils comprise a diverse group of species, each with unique characteristics and ecological roles. Some are notorious agricultural pests, while others are lesser-known contributors to ecosystem dynamics. This chapter highlights key Oriental weevil species, their defining traits, and their significance to both nature and humanity.

- **Key Species Overview**

1. Banana Weevil (Cosmopolites sordidus)

- **Description**: a little weevil that is between 10 and 15 mm long and dark brown or black in color.
- **Habitat**: Primarily found in banana plantations across tropical Asia, Africa, and the Pacific.

Behavior and Impact:
- Lays eggs in the base of banana plants.
- Larvae tunnel through the pseudostem and roots, causing structural damage and reducing yields.
- Considered one of the most destructive pests of bananas globally.

Management: Effective control measures include crop rotation, proper sanitation, and biological agents like parasitic nematodes.

2. Rice Weevil (Sitophilus oryzae)

- **Description**: The elytra (wing covers) of this little, reddish-brown weevil have four tiny yellow dots on them.
- **Habitat**: frequently present in grains that have been kept, particularly corn, wheat, and rice.

Behavior and Impact:

- Infests grains during storage, reducing quality and weight.
- Females bore into grains to lay eggs; larvae develop within, hollowing out the grain.

Management: Fumigation, airtight storage, and natural predators like parasitic wasps help control populations.

3. Sweet Potato Weevil (Cylas formicarius)

- **Description**: A slender, ant-like weevil with a metallic blue or green body and orange-red head.
- **Habitat**: Located in tropical and subtropical areas where sweet potatoes are grown.

Behavior and Impact:

- Larvae seriously harm the tubers by feeding on them.
- Adults consume foliage and tendrils, hence diminishing agricultural yield.

Management: Use of pheromone traps and resistant sweet potato varieties has proven effective.

4. Coconut Rhinoceros Beetle (Oryctes rhinoceros)

(Note: This beetle is often mistaken for a weevil due to similar feeding habits.)

- **Description**: A large black or brown beetle with a distinctive horn-like structure.
- **Habitat**: Found in coconut plantations and palm-growing regions.

Behavior and Impact:

- Adults penetrate the crowns of palms, consuming sap and hindering growth.
- While not a true weevil, its behavior often overlaps with Oriental weevil pests.
- **Defining Characteristics of Each**

These species share common traits such as a preference for plant-based diets and the use of their rostrum for boring into plant material. However, each has unique adaptations suited to its preferred host plants and habitats. For example:

- The banana weevil specializes in tunneling through pseudostems, while the rice weevil focuses on stored grains.

- The sweet potato weevil has evolved to mimic ants for protection, a strategy not seen in other Oriental weevils.

The Lesser-Known Beneficial Species
Not all Oriental weevils are harmful. Some contribute positively to ecosystems by:

1. Pollinating specific tropical plants.
2. Breaking down decaying plant material, aiding nutrient recycling.

3. Acting as a food source for predators, supporting biodiversity.

Examples include species that inhabit forested areas, feeding on wild plants without causing agricultural damage.

Final thought
Understanding the diversity of Oriental weevils is critical for balancing their ecological roles with their impacts on agriculture. By studying both their harmful and beneficial species, we can develop targeted strategies for pest management while preserving the ecological functions they perform.

Part II: Habitat and Distribution

Chapter 5: Habitat and Distribution

Introduction

The distribution of Oriental weevils is vast, extending across tropical and subtropical regions of Asia, Africa, and the Pacific. Their adaptability to various climates, environments, and habitats makes them one of the most widespread groups of insects. Understanding where Oriental weevils thrive and the environmental conditions that support their populations is essential for managing both their beneficial and harmful impacts.

This chapter explores the different habitats that Oriental weevils occupy, their geographic distribution, and the factors that influence their spread.

- Preferred Habitats of Oriental Weevils

1. Tropical and Subtropical Regions
Oriental weevils are predominantly found in tropical and subtropical climates, where the warm temperatures and high humidity support their life cycles. These regions provide ideal conditions for weevils to flourish, particularly in:

- **Rainforests and Tropical Woodlands**: Dense vegetation and rich biodiversity offer a wealth of food sources and shelter.
- **Agricultural Areas**: Crops such as bananas, sugarcane, yams, and rice are commonly affected by weevil infestations.
- **Coastal and Island Environments**: These areas provide unique microhabitats, especially for species like the coconut rhinoceros beetle (Oryctes rhinoceros).

2. Agricultural Land
Oriental weevils are notorious for invading agricultural fields, where they can cause significant damage to crops. These insects typically favor:

- **Root Crops:** Weevils like the banana weevil and sweet potato weevil target root systems and underground plant parts, causing significant crop losses.

- **Stored Products**: Species such as the rice weevil infest stored grains. leading to economic losses in food supply chains.
- **Crops with Soft, Vulnerable Tissue:** Many weevil species are drawn to crops with soft stems or leaves, which they can easily penetrate using their specialized rostrums.

- ## Geographic Distribution of Oriental Weevils

1. Asia
The heart of Oriental weevil diversity lies in Asia, where the warm, humid conditions create an ideal environment for these insects.

- **Southeast Asia**: Countries like Indonesia, the Philippines, and Thailand experience frequent weevil infestations, particularly in banana and coconut plantations.
- **South Asia**: India, Bangladesh, and Sri Lanka host a variety of weevil species that target staple crops like rice and sweet potatoes.

2. Africa

Many Oriental weevils have spread to African countries, especially those with tropical climates like Uganda, Kenya, and Ghana. The banana weevil and the rice weevil have become significant agricultural pests in parts of East and West Africa.

3. Pacific Islands

Island nations in the Pacific, such as Fiji, Samoa, and Hawaii, are often home to both native and invasive Oriental weevil species. The coconut rhinoceros beetle, for example, has had a dramatic impact on coconut plantations in the Pacific, especially in Hawaii and the Philippines.

- **Factors Affecting Distribution**

1. Climate

Temperature and humidity are critical factors in determining where Oriental weevils can thrive. Tropical regions with year-round warmth and moisture support weevil populations, while colder climates may limit their range.

- **Temperature Sensitivity**: Most Oriental weevil species are sensitive to temperature extremes. They thrive in areas with consistently warm conditions but struggle in cooler climates.

- **Rainfall:** High rainfall creates humid environments that are conducive to the life cycles of many weevil species.

2. Human Activity

Human agricultural practices have played a major role in the spread of Oriental weevils.

- **International Trade:** The movement of crops and agricultural products has inadvertently spread weevils to new regions. For example, rice weevils are commonly transported through the global rice trade.
- **Introduction to New Habitats**: Species like the coconut rhinoceros beetle have been introduced to non-native regions through international plant shipments and agricultural expansion.

3. Natural Spread

Weevils are also capable of expanding their range naturally. They can migrate via wind currents or through the movement of host plants.

- **Dispersal Mechanisms**: Some weevil species have evolved the ability to fly or travel via plant material, which aids their natural distribution.

- **Impact of Habitat on Weevil Behavior**

The habitats that Oriental weevils occupy directly influence their behavior, lifecycle, and interaction with plants. For instance:

- **Agricultural Fields:** Weevil species in these environments often exhibit rapid reproductive cycles, enabling them to quickly infest crops and cause damage.
- **Rainforests and Forests**: Weevils in these habitats tend to have slower life cycles but play significant roles in nutrient cycling by decomposing organic material.

Understanding these interactions allows us to better manage weevil populations, preventing overgrowth in agricultural settings while conserving their roles in natural habitats.

Chapter 6: Life Cycle and Behavior

Introduction

The life cycle and behavior of Oriental weevils are crucial factors in understanding their ecological impact, pest status, and the most effective ways to manage them. These insects undergo a complete metamorphosis, transitioning through several life stages, each with distinct behaviors that are shaped by their environment and food sources. In this chapter, we will explore the various stages of the Oriental weevil life cycle, the behavioral patterns that define them, and how these characteristics influence their ability to thrive in different habitats.

• The Life Cycle of Oriental Weevils

Like other beetles, Oriental weevils undergo complete metamorphosis, passing through four main life stages: egg, larva, pupa, and adult. Each stage has its own distinct behavior and environmental requirements.

1. Egg Stage

- **Duration**: Depending on species and environmental conditions, the egg stage typically lasts from 1 to 2 weeks.

Characteristics:

- Eggs are usually laid in or near food sources, such as plant stems, roots, or stored grain.
- Female weevils use their rostrums to bore into plant tissues or fruit to deposit eggs.
- The egg is small, oval, and white or pale in color.

2. Larval Stage

- **Duration:** The larval stage can last anywhere from several weeks to months, depending on species and environmental conditions.

Characteristics:

- Upon hatching, the larvae begin feeding on the plant material, causing damage.
- They are typically C-shaped, white or cream-colored, and legless.
- In species like the banana weevil, larvae feed on the inner tissues of the plant, boring tunnels that can weaken the plant's structural integrity.
- Larvae are particularly damaging to crops because they feed on roots, stems, or tubers, directly affecting plant health and growth.

3. Pupal Stage

- **Duration:** The pupal stage usually lasts between 2 to 3 weeks.

Characteristics:

- The pupa is typically found in the soil or within plant material, depending on the species.
- During this stage, the larvae undergo a transformation into adults.

- Pupal cases can be hard and protective, offering the insect a safe transition to adulthood.
- It is in this stage that weevils develop their characteristic rostrum and hardened exoskeleton.

4. Adult Stage

- **Duration:** Adult weevils can live for several months, sometimes even up to a year, depending on food availability and environmental conditions.

Characteristics:

- Adult Oriental weevils are generally small to medium-sized, with hard, protective exoskeletons and distinctive long rostrums.
- Adults are capable of flight, though some species are flightless.
- They feed on leaves, stems, roots, or fruit, depending on their diet preferences.
- Mating occurs during this stage, with females laying eggs in suitable locations to continue the life cycle.
- Many species exhibit nocturnal activity patterns, with adults feeding primarily at night.

Weevil Behavior

Behavioral patterns in Oriental weevils vary between species but are typically influenced by their environment and their need to feed and reproduce. Understanding these behaviors is crucial for pest management strategies and ecosystem studies.

1. Feeding Behavior

Weevils are predominantly herbivorous, and their feeding behavior is often linked to plant damage:

- **Root and Stem Boring:** Many Oriental weevils, such as the banana weevil and sweet potato weevil, bore into the roots and stems of their host plants, causing significant structural damage.
- **Grain Infestation**: Species like the rice weevil infest stored grains, where adults and larvae consume the grain from the inside, making it unfit for consumption.
- **Host Plant Selection:** Weevils are often host-specific, meaning they tend to favor certain plants. This selective feeding behavior can lead to localized damage in crops that are their preferred hosts.

2. Mating and Reproduction

Mating behavior in Oriental weevils is typically solitary, with males seeking out females through pheromones or physical cues.

- **Mating Rituals:** In some species, males may exhibit courtship behaviors, such as antennal tapping or subtle body movements, to attract mates.
- **Egg-Laying:** Females use their rostrums to puncture plant tissues, creating a suitable place for egg deposition. They typically lay eggs in groups or singly, depending on the species.

3. Flight and Dispersal

Many Oriental weevil species are capable of flight, although the distance they can travel varies by species.

- **Dispersal Mechanisms:** Adult weevils often fly short distances in search of food or suitable mating sites.
- **Migration Patterns**: In some cases, weevils may move to new habitats or crops, either due to resource depletion or environmental factors like temperature and humidity changes.

4. Nocturnal Behavior

Many Oriental weevil species are nocturnal, feeding and mating primarily at night. This behavior is likely an adaptation to avoid predators and environmental extremes.

- **Light Sensitivity**: Weevils are often attracted to artificial lights, which can help in monitoring their activity patterns.

Final thought

The life cycle and behavior of Oriental weevils are intricately tied to their ecological niches and agricultural impact. By understanding these patterns, we can better predict and manage their populations, especially in regions where they pose a threat to crops. Furthermore, studying weevil behavior contributes to our broader understanding of insect physiology, reproductive strategies, and ecological roles. Through continued research, we can improve pest management practices and reduce the adverse effects of these fascinating insects.

Part III: Ecology and Environmental Role

Chapter 7: Interaction with Host Plants

Introduction

The relationship between Oriental weevils and their host plants is fundamental to understanding their ecological roles and agricultural impacts. These interactions often shape the behavior and life cycles of the weevils, as they are primarily herbivorous, feeding on plant tissues such as stems, roots, and leaves. Weevil larvae and adults alike contribute to plant damage, sometimes leading to significant crop losses. However, these interactions also highlight the weevil's ecological significance, as they often influence plant health, competition, and biodiversity.

In this chapter, we explore how Oriental weevils interact with their host plants, focusing on their feeding patterns, reproductive behaviors, and the resultant effects on plant physiology. Additionally, we'll discuss the mechanisms through which weevils select host plants and the broader ecological implications of these interactions.

- ### Feeding Mechanisms of Oriental Weevils

1. Root and Stem Feeding

A key characteristic of many Oriental weevils is their ability to bore into the tissues of their host plants. These weevils typically target plant roots, stems, or tubers. The damage caused can be severe, weakening the plant and sometimes causing death.

- **Banana Weevil (Cosmopolites sordidus**): This species is notorious for feeding on the pseudostem of banana plants. The larvae tunnel through the stem, causing structural damage that makes the plant vulnerable to wind and water stress.
- **Sweet Potato Weevil (Cylas formicarius):** Larvae of this species feed on the tubers of sweet potatoes, hollowing them out and reducing their quality. This feeding behavior often results in crop failures.
- **Rice Weevil (Sitophilus oryzae**): While this species is more often found in stored grains than in live plants, it has a similar feeding strategy, as it bores into grains, leaving them hollow and unfit for consumption.

These feeding behaviors disrupt the plant's vascular system, making it difficult for plants to absorb nutrients and water. This can lead to stunted growth, reduced yield, or even plant death.

2. Leaf and Fruit Feeding

While less common, some Oriental weevils also feed on the leaves or fruits of plants, which can affect photosynthesis and fruit production.

- **Coconut Rhinoceros Beetle (Oryctes rhinoceros):** This species feeds on the soft tissue at the tops of coconut palms. By chewing the growing tissue, it stunts the palm's growth and reduces its ability to produce coconuts.
- **Other Weevil Species**: Certain weevils are known to feed on the leaves of various plants, causing defoliation. In the case of ornamental plants or crops like soybeans and peanuts, this can significantly reduce productivity.

- ## Host Plant Selection and Preferences

1. Plant Species Preferences

Weevils are generally host-specific, meaning that each species of Oriental weevil tends to favor certain plants for feeding and reproduction. The selection of host plants is influenced by factors such as:

- **Chemical Composition**: Some plants produce chemicals (like alkaloids, tannins, or resins) that act as deterrents to certain weevils but attract others.
- **Nutrient Content:** Plants with higher levels of starch, sugar, or protein are often more attractive to weevils, which require these nutrients for reproduction and development.

2. Plant Size and Condition

The size and condition of the plant also influence a weevil's choice of feeding site. Larger plants with more accessible root systems or soft tissue are more likely to be attacked, while younger or weaker plants are more vulnerable to damage. This preference allows weevils to maximize their chances of survival and reproduction.

3. Ecological Impact

By feeding on specific host plants, weevils can shape the plant community in their environment.

- **Competition:** Weevil feeding can weaken or kill plants, allowing other species to dominate. This can alter species diversity, particularly in agricultural settings.
- **Evolutionary Pressure**: Plants may evolve to resist weevil attacks through physical or chemical defenses. This leads to coevolution between plants and their insect pests, driving adaptations in both species.

- ## Reproductive Behavior and Plant Interaction

1. Egg Laying
The way weevils lay their eggs is intimately tied to the plant material they feed on. Females will often bore into the stems or roots of plants using their rostrum, creating a safe environment for their eggs.

- **Banana Weevil**: Females insert eggs into the base of the banana plant, where the larvae will hatch and immediately begin feeding on the plant tissues.
- **Sweet Potato Weevil**: Eggs are often laid directly into the sweet potato tubers, and the larvae consume the tubers as they develop.

The choice of egg-laying site is crucial to the survival of the larvae, as it ensures that they have an immediate food source upon hatching. This strategy often results in localized infestations, where a single weevil can cause significant damage to a crop or plant population.

2. Larval Development and Plant Impact

The larvae keep eating the host plant when they hatch. As they mature, their feeding can cause extensive damage, leading to:

- **Reduced Photosynthesis**: Damage to leaves and stems limits the plant's ability to conduct photosynthesis, stunting growth.
- **Weakened Structure:** Weevil larvae often tunnel into the plant's vascular system, weakening the plant's structural integrity and making it more susceptible to diseases and environmental stress.

• Weevil-Plant Coevolution

The long-standing relationship between weevils and their host plants has led to a process of coevolution, where both species exert selective pressures on one another.

- **Defensive Adaptations Plants**: Some plants have evolved physical defenses like tough outer skins, thorns, or chemical compounds that deter weevils.

- **Behavioral Adaptations in Weevils**: In turn, weevils have developed specialized feeding structures and behaviors to overcome these defenses, such as stronger rostrums for boring through tough plant material.

This coevolutionary dynamic shapes the nature of plant-insect interactions, influencing biodiversity and ecosystem stability.

Final thought

The interaction between Oriental weevils and their host plants is a complex and dynamic relationship that is essential to both their life cycle and ecological roles. Weevils contribute to plant damage, particularly in agricultural settings, but they also play a significant role in shaping plant communities and driving evolutionary processes.

Chapter 8: Ecological Role of Oriental Weevils

Introduction

While Oriental weevils are often considered agricultural pests due to their feeding habits and impact on crops, they also play crucial roles in the ecosystems they inhabit. These insects contribute to nutrient cycling, plant diversity, and even serve as prey for a range of predators. Understanding the ecological role of Oriental weevils is essential not only for pest management but also for appreciating their place in broader ecological processes.

In this chapter, we explore the diverse roles Oriental weevils play in their environments, from their influence on plant health and decomposition to their interactions with other species in food webs. By examining both their positive and negative ecological impacts, we can better understand how to balance their role as pests and as integral components of ecosystems.

- ## Nutrient Cycling and Decomposition

One of the most important ecological functions of Oriental weevils is their role in nutrient cycling. Through their feeding and burrowing behaviors, weevils contribute to the breakdown of plant material, which in turn supports the regeneration of soil nutrients.

1. Decomposition Process

Many Oriental weevils, especially those that feed on plant roots, stems, and decaying plant matter, contribute to the decomposition process.

- **Breaking Down Organic Matter**: As weevils feed on dead or decaying plant material, they help break it down into smaller pieces, facilitating the activity of microorganisms that decompose organic matter.
- **Soil Enrichment**: The waste produced by weevils, as well as the decomposing plant material, enriches the soil with essential nutrients like nitrogen and phosphorus. This process supports the growth of new vegetation and maintains soil fertility.

2. Contribution to Soil Structure

Weevils that burrow into soil or plant tissues, such as the larvae of root-feeding species, also help in the aeration of the soil.

- **Soil Aeration**: Weevil tunnels create pathways that allow air and water to penetrate deeper into the soil, improving the oxygen availability for plant roots and microorganisms.

- **Soil Mix and Compaction Reduction**: Their burrowing activity can reduce soil compaction, promoting better root growth and improving plant health in the long term.

• Weevils as Prey for Predators

Weevils are an important food source for a variety of predators, playing a critical role in the food web. From birds to mammals and other insects, many species depend on weevils as a primary or secondary food source.

1. Predator Species

Weevils, especially their larvae, serve as a nutritious food source for numerous animals:

- **Birds**: Many birds, particularly those that forage in agricultural fields or forests, consume weevils and their larvae.
- **Mammals**: Small mammals such as rodents are known to feed on weevils, particularly in areas where crops or plant material are abundant.
- **Other Insects**: Certain predatory insects, such as ants, spiders, and beetles, feed on adult weevils and their larvae.

Weevil populations can thus support a wide range of predators, helping to maintain balance within the ecosystem. In areas where weevil populations are high, they often become a key component of local food chains.

• Weevils and Plant Diversity

Although many weevil species can damage crops and ornamental plants, their interactions with wild plant species have broader ecological implications. Through their feeding habits, weevils can influence plant diversity and ecosystem structure in natural habitats.

1. Plant Competition and Community Dynamics

Weevils can indirectly influence plant diversity by feeding on dominant species and giving less competitive plants an opportunity to thrive.

- **Weed Suppression**: In some environments, weevil species that feed on invasive or dominant plant species can help limit their spread, allowing native plants to grow and maintain biodiversity.
- **Reduction of Weevil-Sensitive Species:** Conversely, weevils can harm certain plant species, especially those that are more vulnerable to their feeding. This can lead to

changes in plant community composition, where weevil-resistant species become more dominant.

2. Pollination and Plant Interaction

Some weevils are known to play a role in plant pollination. Although not as efficient as bees or butterflies, certain weevil species can contribute to pollination, especially in tropical ecosystems where they visit flowers while feeding on plant tissues.

- ## Weevils in Agricultural Systems

In agricultural systems, Oriental weevils are often categorized as pests, but their role in ecosystems is more nuanced. While they can cause crop damage, they also contribute to the ecological health of the environment in which they live.

1. Impact on Agricultural Biodiversity

In areas where monoculture farming is common, weevils may disrupt crop health by targeting specific plant species. However, their presence can also stimulate biodiversity by promoting the growth of pest-resistant plant varieties.

- **Crop Rotation and Biodiversity**: In agricultural systems that practice crop rotation, weevils help maintain soil health and may reduce pest pressures on certain crops.

- **Natural Pest Control**: Weevils can sometimes aid in controlling other pests by feeding on them or reducing the dominance of certain plant species that might harbor other harmful insects.

2. Balancing the Agricultural Ecosystem

Farmers often rely on chemical pesticides to control weevil populations, but this can have detrimental effects on the surrounding environment. By understanding the ecological role of weevils, farmers can adopt integrated pest management (IPM) strategies that focus on balancing pest control with ecological health.

- **IPM Practices:** Strategies like introducing natural predators (such as parasitoid wasps) or cultivating pest-resistant crops can help control weevil populations without disrupting the entire ecosystem

- **Ecological and Agricultural Management**

Understanding the ecological role of Oriental weevils helps inform better management practices. By considering their contributions to soil health, biodiversity, and predator-prey dynamics, we can develop more sustainable pest control strategies.

- **Ecosystem Restoration**: In some regions, reintroducing weevil populations to degraded ecosystems may help restore soil health and biodiversity.
- **Conservation of Beneficial Species:** Ensuring that beneficial weevil species, which aid in decomposition and nutrient cycling, are preserved is vital for maintaining ecosystem balance.

Part Iv: Anatomy and Biology

To manage Oriental weevils effectively, one must understand their anatomy and biology. These pests possess unique physical and behavioral traits that contribute to their adaptability and impact. This section explores the weevil's anatomy, life cycle, and reproductive strategies in detail.

Chapter 9: Anatomy of a Weevil

External Features: Rostrum, Antennae, and Legs

1.Rostrum (Snout)
- The rostrum is a defining characteristic of weevils. It is elongated and equipped with specialized mouthparts at the tip, used to pierce and chew plant tissues.
- In Oriental weevils, the rostrum also serves as a tool for females to bore into plants for

egg-laying, ensuring the larvae are born in a food-rich environment.

2. Antennae

- The antennae are jointed and often bent like elbows (geniculate). Positioned on the rostrum, they are sensitive to chemical signals, helping weevils locate food, mates, and suitable sites for egg-laying.
- Their clubbed ends enhance their efficiency in sensing environmental changes.

3. Legs

- Weevils have three pairs of legs, each adapted for gripping surfaces.
- Their legs are robust, enabling them to traverse various terrains, from smooth leaves to rugged soil.

Internal Systems

1. Digestive System

- A highly specialized system for digesting tough plant materials.
- Includes salivary enzymes that break down cellulose and starch, aiding in efficient nutrient extraction.

2. Respiratory System

- Weevils rely on a network of tracheae and spiracles for breathing. These structures ensure oxygen reaches internal tissues efficiently.

3. Nervous System
- A central ganglion controls motor and sensory functions, allowing precise movements and responses to environmental stimuli.

4. Reproductive System
- Females have specialized organs for egg production, capable of laying hundreds of eggs during their lifetime.
- Males possess structures optimized for mating, ensuring effective fertilization.

Life Cycle and Reproduction

Egg, Larva, Pupa, and Adult Stages

1.Egg Stage
- Oriental weevils deposit their eggs in the stems, seeds, or roots of plants. This hidden placement protects the eggs from predators and environmental stress.

- Depending on the environment, the small, oval, and pale eggs might take anywhere from a few days to several weeks to incubate.

2. Larva Stage
- Larvae are soft-bodied, legless, and highly destructive. They burrow into plant tissues, feeding on roots, seeds, or stems, causing significant agricultural damage.
- This stage lasts for several weeks, during which larvae molt multiple times as they grow.

3. Pupa Stage
- The pupa is the transitional stage where the larva transforms into an adult.
- Oriental weevils often pupate within the plant tissues or soil, encased in a protective cocoon-like structure. This stage is critical for their metamorphosis.

4. Adult Stage
- Adult weevils emerge fully formed with hardened exoskeletons and functional rostra.
- They disperse to new locations, feeding, mating, and laying eggs to complete the cycle.

Reproductive Behavior and Strategies

1.Mating Behavior

- Mating is preceded by courtship, where males may display specific behaviors to attract females.
- Mating can last for hours, ensuring that fertilization is successful.

2. Egg-Laying Strategies

- Females select optimal plant hosts, boring small holes in plant tissues to deposit eggs. This strategic placement provides the emerging larvae with an immediate food source.

3. Generation Overlap

- Oriental weevils often produce overlapping generations, with adults, larvae, and eggs coexisting in the same environment. This continuous life cycle exacerbates their impact on crops.

Final thought

The intricate anatomy and biology of Oriental weevils reveal their evolutionary success as pests. Their specialized structures, diverse life stages, and reproductive strategies make them resilient and challenging to control. By targeting vulnerable points in their anatomy and life cycle, pest management strategies can be developed to curb their spread and minimize their damage effectively.

Part V: Management and Control

Chapter 10: Management and Control Methods

Introduction

Managing Oriental weevils is crucial in agricultural systems and natural ecosystems where their feeding can lead to significant crop losses and ecological disruptions. Effective management strategies must balance pest control with environmental considerations to minimize harm to beneficial species and avoid excessive chemical use. This chapter explores the various methods available for controlling Oriental weevil populations, focusing on both traditional and modern approaches, including chemical, biological, cultural, and integrated pest management (IPM) techniques.

- ## Chemical Control Methods

Chemical control remains one of the most commonly used methods for managing Oriental weevils, particularly in agriculture. However, its effectiveness can vary, and there are concerns regarding environmental impact, resistance development, and non-target effects.

1. Insecticides

Insecticides are widely used to target both adult and larval weevils. Several classes of insecticides are effective against weevils, including:

- **Organophosphates**: These are effective in controlling both larvae and adults. However, their toxicity to beneficial insects and the potential for resistance development have limited their use in some regions.
- **Pyrethroids:** These synthetic insecticides are widely used because of their broad-spectrum effectiveness against weevils. However, overuse can lead to resistance and harm to pollinators.
- **Neonicotinoids:** These are systemic insecticides that are absorbed by plants and target insect pests when they feed on the plant. They can be effective against weevils, especially in stored grains, but they also pose risks to non-target insects, including pollinators.

2. Application Techniques

To maximize the effectiveness of chemical control, proper application methods are critical:

- **Soil Drenching:** For root-feeding weevils, insecticides may be applied directly to the soil to target larvae.
- **Foliar Sprays:** In cases of adult weevils feeding on plant leaves, insecticides are sprayed directly onto the foliage.
- **Granular Insecticides**: These are applied to the soil or base of plants, where they are taken up by weevil larvae as they feed on plant roots.

While chemical methods are effective, they should be used with caution to avoid environmental harm and resistance development.

- ## Biological Control Methods

Biological control offers an eco-friendly alternative to chemical pesticides by introducing natural predators or pathogens that target Oriental weevil populations. This approach can be highly effective when used in combination with other methods, such as cultural practices or integrated pest management.

1. Natural Predators

Several natural predators, including parasitoid wasps and beetles, can help control weevil populations by targeting eggs, larvae, and adults.

- **Parasitoid Wasps**: These tiny wasps lay their eggs inside weevil larvae, where their offspring feed on and eventually kill the host. The introduction of parasitoid wasps can significantly reduce weevil populations in controlled environments.
- **Predatory Beetles**: Certain beetles feed on weevil larvae, helping to reduce pest numbers. These beetles are particularly effective in forests and natural habitats where weevil populations can become concentrated.

2. Microbial Control

Microorganisms, including bacteria, fungi, and viruses, can be used to target Oriental weevils at different life stages.

- **Entomopathogenic Fungi**: Fungi such as Beauveria bassiana and Metarhizium anisopliae are pathogenic to weevils and can be used as a biological control. When applied to affected crops or soil, these fungi infect and kill weevils.

- **Bacterial Insecticides**: Bacillus thuringiensis (Bt), a bacterium that produces toxins harmful to insects, has shown effectiveness against some weevil species, particularly larvae.
- **Nematodes:** Certain nematodes, such as Steinernema and Heterorhabditis species, parasitize weevil larvae by entering their bodies and killing them.

Biological control methods are often slower to show results than chemical methods, but they offer long-term solutions without harming the environment.

• Cultural Control Methods

Cultural control refers to farming practices that reduce weevil populations by altering the environment to make it less favorable for weevil survival and reproduction. These methods focus on prevention and are often used in combination with other control strategies.

1. Crop Rotation
By alternating the types of crops grown in a given field, farmers can disrupt the life cycle of weevils that are host-specific. This reduces the availability of food for weevils and limits their ability to build up large populations in a single crop.

- **Rotation with Non-Host Crops**: Planting crops that are not suitable for weevil feeding can significantly reduce weevil numbers in the soil.
- **Break Crops**: Certain crops, such as legumes or mustard, are not attractive to weevils and can help break the cycle of infestation.

2. Resistant Varieties

Growing weevil-resistant plant varieties can help mitigate the damage caused by these pests. Through selective breeding or genetic modification, some crops have been developed to tolerate or repel weevil infestations.

- **Genetically Modified Crops**: Some genetically modified plants are resistant to weevil damage, either by producing toxins that deter feeding or by being less susceptible to the larvae's tunneling.
- **Naturally Resistant Varieties**: Breeding for naturally resistant varieties can reduce reliance on chemical pesticides and improve overall crop health.

3. Sanitation Practices

Proper field sanitation can also help reduce weevil populations. This includes removing plant debris, infested crops, and weeds, all of which may serve as breeding grounds for weevils.

• Integrated Pest Management (IPM)

Integrated Pest Management (IPM) is an approach that combines multiple control methods—chemical, biological, cultural, and mechanical—to manage weevil populations in a way that is environmentally sustainable and economically viable. The key to a successful IPM program is monitoring, early detection, and taking action before infestations reach damaging levels.

1. Monitoring and Early Detection
Regular monitoring of crops and plants for signs of weevil infestation is essential. This includes using traps, visual inspections, and soil sampling to detect the presence of adult weevils and larvae. Early detection helps to implement control measures before populations become large and difficult to manage.

2. Threshold Levels
IPM relies on setting threshold levels for pest damage. These thresholds indicate when intervention is needed to prevent economic loss or

significant ecological harm. For Oriental weevils, this might mean applying a control measure once a certain percentage of plants are infested or damage reaches a certain level.

3. Combination of Methods
In an IPM strategy, a combination of methods is used to manage weevil populations. For example, biological control agents like parasitoid wasps might be introduced in conjunction with soil drenching of insecticides or the use of resistant plant varieties. This approach minimizes the impact of any one method while maximizing effectiveness.

Chapter 12: Case Studies on Weevil Management

Introduction
Real-world case studies provide invaluable insight into how Oriental weevil management strategies are applied in diverse environments. By examining successful and unsuccessful examples of pest control, we can learn about the challenges, solutions, and practical applications of different methods. This chapter highlights a selection of case studies from agriculture, forestry, and urban

settings, demonstrating various approaches to managing weevil populations effectively.

These case studies not only illustrate the complexity of weevil management but also provide lessons on integrating pest control strategies, addressing environmental concerns, and adjusting practices based on specific ecological conditions. By analyzing these real-life examples, we can better understand the multifaceted nature of weevil control and the importance of adaptive management strategies.

Case Study 1: Agricultural Management of Rice Weevils in Southeast Asia

1. Background

Rice weevils (Sitophilus oryzae), a major pest in Southeast Asia, affect rice storage and cause significant post-harvest losses. These weevils attack stored rice grains, leading to reduced quality and market value. In response, Southeast Asian countries have employed a combination of chemical, biological, and cultural control methods to manage weevil populations.

2. Management Approach
- **Chemical Control**: Farmers initially relied heavily on synthetic insecticides to treat

stored rice. However, this led to concerns about chemical residues in food products and resistance development in weevil populations.

- **Biological Control:** Researchers introduced the parasitoid Anisopteromalus calandrae, a natural enemy of the rice weevil, into rice storage facilities. This parasitoid was shown to reduce weevil populations significantly without harming the quality of the rice.
- **Cultural Control:** Farmers began adopting better post-harvest handling practices, including cleaning storage facilities, using airtight containers, and applying controlled temperatures to limit weevil reproduction.

3. Outcome

The integrated use of biological and cultural methods proved successful in reducing rice weevil populations and post-harvest losses. The combination of these approaches also reduced pesticide use, improving food safety and sustainability. Over time, these practices became standard in many regions, highlighting the effectiveness of IPM in managing weevils.

Case Study 2: Control of the Red Palm Weevil in the Middle East

1. Background

The red palm weevil (Rhynchophorus ferrugineus) is a devastating pest of palm trees, particularly date palms, in the Middle East. It causes significant damage to palm plantations, leading to the destruction of the trees and affecting the livelihoods of local farmers. The red palm weevil burrows into the trunk of the palm tree, causing weakening and eventual death of the plant.

2. Management Approach

- **Chemical Control:** Chemical insecticides were initially used to combat the weevil, but the widespread application led to concerns about resistance and environmental contamination.
- **Biological Control**: In response, researchers introduced natural enemies, such as the parasitic wasp Baryscapus servadeii, which targets red palm weevil larvae. These parasitoids significantly reduced weevil populations when used alongside monitoring systems to detect infestations early.
- **Cultural Control**: Farmers implemented more rigorous tree inspections, early removal of infested palms, and the use of pheromone traps to detect weevil presence before damage occurred. In addition, the pruning of palm trees and regular sanitation

of plantations helped limit the spread of the pest.

3. Outcome

The combination of pheromone traps, parasitoids, and cultural practices led to a reduction in red palm weevil populations and improved palm health. Though chemical treatments were still used in severe infestations, the overall reliance on pesticides decreased, contributing to better environmental outcomes and lower costs for farmers.

Case Study 3: Forestry Management of the Pine Weevil in Europe

1. Background

The pine weevil (Hylobius abietis) is a major pest of newly planted pine trees in Europe. These weevils damage seedlings by feeding on the bark, leading to tree mortality and reduced forest regeneration. Due to their impact on forestry, management strategies have been developed to prevent weevil damage to young trees.

2. Management Approach

- **Chemical Control**: The use of insecticides, particularly pyrethroids, has been a common method for controlling pine weevil populations. However, this approach raises concerns about non-target effects, such as harm to beneficial insects and soil organisms.

- **Biological Control**: The introduction of Beauveria bassiana, a fungal pathogen that targets pine weevil larvae, provided a more sustainable alternative to chemical control. Fungal applications have been shown to reduce weevil larvae numbers significantly without harming the environment.
- **Cultural Control**: Forestry managers implemented preventive measures such as the use of protective bark sprays on young trees and the careful timing of planting to avoid peak weevil activity. Rotating the species of trees planted in infested areas also helped reduce weevil populations by disrupting their life cycle.

3. Outcome

The combination of biological control and cultural methods, including the application of fungal agents and the careful management of planting schedules, resulted in a significant reduction in pine weevil damage. Although chemical control was still sometimes used, integrated approaches provided a more sustainable solution for large-scale forestry operations.

Case Study 4: Urban Management of Weevils in Public Parks and Gardens

1. Background

Weevil infestations in public parks and gardens can lead to aesthetic damage and the loss of ornamental plants. In urban areas, managing weevil populations is often a balance between preventing damage and minimizing pesticide use, particularly in spaces where the public interacts with plants.

2. Management Approach

- **Cultural Control**: In urban parks, horticulturists focused on plant health through proper irrigation, mulching, and the use of resistant plant varieties. These practices helped reduce the likelihood of weevil infestations by creating a more robust and less susceptible plant community.
- **Biological Control**: The use of beneficial insects such as predatory beetles and parasitic wasps was explored as a way to control weevil populations in public gardens. These natural predators were introduced to target weevil larvae without harming other beneficial species.

- **Chemical Control**: In extreme cases, targeted pesticide applications were used in affected areas. However, care was taken to select low-toxicity products and apply them only when necessary to minimize harm to non-target species.

3. Outcome

The combination of cultural practices and biological control methods allowed for effective management of weevil populations without widespread pesticide use. Public awareness campaigns also helped gardeners and park maintenance teams recognize early signs of infestation, leading to quicker intervention and less damage.

Final thought

These case studies highlight the diverse approaches to Oriental weevil management across various environments. From agriculture to urban parks, successful management strategies often rely on a combination of methods that are tailored to the specific ecological and economic conditions of each situation. The integration of chemical, biological, and cultural controls, as well as the adoption of IPM practices, has proven to be the most effective way to manage weevil populations while minimizing environmental impact.

Part VI: Future Perspectives

Chapter 12: Future Challenges and Directions in Weevil Research

Introduction

Weevil species, including the Oriental weevil, continue to pose significant challenges in agriculture, forestry, and natural ecosystems. As global trade and climate change continue to affect pest distributions, we must adapt our management strategies to address these evolving challenges. Ongoing research is crucial for understanding weevil behavior, biology, and ecology, which will inform the development of more effective and sustainable pest control methods.

This chapter explores the key challenges faced by researchers and pest managers when dealing with Oriental weevils and outlines the future directions for weevil research. By addressing these emerging issues, we can improve our ability to control these pests and mitigate their impacts on global food security and biodiversity.

1. Climate Change and Weevil Distribution

1. Impact of Global Warming on Weevil Populations

Climate change is expected to have a significant impact on the distribution and behavior of Oriental weevils. Warmer temperatures may enable weevils to expand their range into new regions, potentially affecting crops and ecosystems in areas previously unaffected. Changes in precipitation patterns may also influence weevil lifecycle timing, leading to shifts in infestation peaks and increased pest pressure in new environments.

2. Predicting Future Spread

One of the key challenges in weevil research is developing reliable models to predict how climate change will affect weevil populations. Research on the relationship between temperature, humidity, and weevil reproduction will help scientists develop models to forecast where new infestations may occur. This could enable preemptive management strategies and the allocation of resources to at-risk regions before weevils become established.

3. Adaptation to Climate Change

Weevil species may also adapt to new climatic conditions, possibly leading to the emergence of new pest strains with altered feeding habits or increased resistance to current control methods. Understanding how Oriental weevils adapt to changing environmental conditions will be key in developing long-term management solutions that can keep pace with climate-induced shifts in pest dynamics.

2. Resistance Development in Weevils

1. Insecticide Resistance

One of the biggest challenges in weevil management is the development of resistance to chemical insecticides. Over time, Oriental weevils, like many pest species, have shown the ability to evolve resistance to common insecticides. This reduces the effectiveness of chemical control methods and leads to the need for higher doses or more frequent applications, which can have negative environmental and economic consequences.

2. Addressing Resistance

Future research will need to focus on understanding the mechanisms behind resistance development in Oriental weevils. Studies on genetic mutations, metabolic changes, and behavioral adaptations that confer resistance are critical for developing new insecticides or alternative control strategies. Additionally, researchers are exploring methods such as genetic modification and gene editing to develop weevil populations that are less likely to develop resistance.

3. Alternatives to Chemical Control

To address resistance, there is a growing need to explore alternative control methods, such as biological control agents, novel insecticides, and integrated pest management (IPM) approaches. Advances in biocontrol, such as the use of parasitoid wasps and entomopathogenic fungi, offer promising avenues for reducing reliance on chemicals and preventing resistance buildup.

3. Enhancing Biological Control

1. Expanding the Use of Natural Enemies

Biological control is an environmentally friendly and sustainable approach to managing weevil populations. However, more research is needed to identify new natural enemies that can target Oriental weevils at different life stages.

Additionally, researchers are investigating the mass rearing of parasitoids and predators to ensure their widespread availability for use in pest management programs.

2. Efficacy and Safety of Biocontrol Agents

While biological control agents can be highly effective, ensuring their safety and efficacy in diverse ecosystems is critical. Research into the ecological impact of introducing new species, including potential non-target effects, will help prevent unintended consequences in natural habitats. The success of biocontrol also depends on optimizing the release strategies, timing, and environmental conditions under which these agents perform best.

3. Genetic Engineering of Biocontrol Agents

One emerging area of research involves the genetic modification of biocontrol agents to enhance their ability to target weevils. For example, researchers are exploring the possibility of creating genetically engineered parasitoid wasps that are more efficient at parasitizing weevil larvae. This could help reduce the reliance on pesticides while improving pest control outcomes.

4. Development of Resistant Crop Varieties

1. Breeding for Weevil Resistance

One promising area of research is the development of crop varieties that are resistant to weevil infestations. By identifying and incorporating resistance traits into commercial crops, researchers can help reduce the damage caused by weevils. This could involve genetic modification to enhance plant defenses or traditional breeding methods to select for traits that deter weevil feeding or disrupt their reproductive cycles.

2. Genetic Modification and CRISPR

Advances in genetic engineering, particularly through the use of CRISPR technology, may offer new ways to develop crop varieties that are highly resistant to weevils. By editing the genes of plants to produce compounds that repel or deter weevils, researchers could create crops that are more resilient to pest damage. CRISPR could also be used to target genes that regulate weevil behavior, making it more difficult for them to infest crops.

3. Multi-Trait Resistance

In addition to developing weevil-resistant crops, researchers are also focusing on creating varieties that are resistant to multiple pests and diseases. This approach would reduce the need for frequent pesticide applications and help sustain healthy ecosystems by promoting biodiversity. It also aligns with sustainable agricultural practices by reducing the reliance on chemicals and promoting the use of integrated pest management.

5. Technological Innovations in Weevil Monitoring

1. Smart Traps and Detection Systems

The development of advanced monitoring systems is crucial for early detection and timely intervention in weevil infestations. Researchers are working on "smart traps" that use sensors and artificial intelligence (AI) to monitor weevil populations in real time. These traps can automatically identify weevil species, count their numbers, and even track their movements, providing farmers and pest managers with valuable data to optimize pest control strategies.

2. Remote Sensing and Drones

Remote sensing technologies, such as satellite imagery and drones, are increasingly being used to monitor large-scale agricultural areas for signs of weevil infestations. By collecting data on crop health, temperature, and humidity, drones can help detect weevil activity before visible damage occurs, enabling early intervention. This could lead to more precise pest management and reduce the use of pesticides.

3. Genomic Tools for Weevil Identification
Advancements in genomic sequencing and DNA barcoding are providing researchers with powerful tools to identify and track different weevil species. These tools allow for more accurate species identification, which is essential for developing targeted control strategies. Furthermore, genomic tools can help monitor weevil populations for genetic resistance and inform breeding programs for resistant crop varieties.

6. Public Awareness and Policy Development

1. Building Public Awareness
Increasing public awareness of Oriental weevil threats is essential for encouraging proactive management efforts. Educating farmers, gardeners, and the general public about the risks of weevils and how to recognize and prevent infestations can help limit the spread of these pests. Outreach programs, training workshops, and informational campaigns can play a crucial role in fostering a collective response to weevil management.

2. Policy and Regulatory Support

Governments and international organizations play a critical role in supporting research and establishing policies that guide weevil management. This includes funding for research, the development of pest management guidelines, and regulations regarding the use of pesticides and biocontrol agents. International cooperation is also essential, particularly in areas where weevil species cross national borders.

Final thought

The future of Oriental weevil management lies in a combination of innovative research and adaptive management strategies. Addressing challenges such as climate change, resistance development, and the integration of new technologies will require collaboration across disciplines and regions. By focusing on sustainable, integrated approaches, researchers and pest managers can mitigate the impact of Oriental weevils while protecting agricultural systems and natural ecosystems.

Chapter 13: Conclusion

● Summary of Key Findings

Throughout this book, we've explored the various facets of Oriental weevils, from their biology and ecology to their management and control. Oriental weevils, like many other pest species, have significant implications for agriculture, forestry, and natural ecosystems. Their adaptability and wide distribution highlight the ongoing challenge of managing their populations.

We've seen how various strategies—ranging from chemical and biological control to integrated pest management (IPM)—have been employed with varying degrees of success. Case studies across different regions show that a multifaceted approach, which includes cultural, chemical, and biological strategies, is the most effective in minimizing damage while also protecting the environment and public health.

Challenges in Weevil Management

Weevil management is far from simple. Resistance to insecticides, climate change, and the potential for unintended ecological impacts from control measures pose ongoing challenges. The growing awareness of these issues has led to more sustainable approaches, such as biocontrol and the development of resistant crop varieties.

However, the complexity of pest management means that research must continue to evolve. Scientists must explore new technologies, such as AI for monitoring and CRISPR for crop improvement, to stay ahead of these adaptable pests.

- ### The Importance of Ongoing Research

As the global climate continues to shift and agricultural systems face new pressures, the need for continuous research on weevil biology, control methods, and pest dynamics becomes ever more crucial. Understanding the behavior of Oriental weevils and their interactions with different ecosystems will be key to designing more effective, environmentally-friendly management strategies.

Future Directions

The future of weevil research will likely focus on integrating advanced technologies, improving biological control agents, and developing crops that can resist pest damage. Innovations in genomics, biocontrol, and monitoring systems will provide new tools for pest managers, helping to ensure that weevil populations are controlled with minimal impact on the environment.

Additionally, cross-disciplinary research involving climate scientists, ecologists, agricultural experts, and pest managers will be essential for anticipating and responding to the effects of climate change on weevil distribution.

Final Thoughts
The challenge of controlling Oriental weevils is ongoing, but through sustained research, collaboration, and innovation, we can create more effective and sustainable management solutions. The lessons learned from past efforts, combined with new technological advancements, provide hope for minimizing the impact of these pests on both agriculture and natural ecosystems.

Thank You!
Thank you for taking the time to explore the fascinating world of Oriental weevils with

this book. Your interest in understanding these pests and their impact on agriculture and ecosystems is invaluable. As we continue to learn and adapt to the challenges posed by these resilient creatures, your curiosity and dedication to sustainable practices contribute to a better, more informed world.

I hope this book has sparked new ideas and provided practical insights for managing Oriental weevils. Remember, the future of pest management lies in innovation, collaboration, and an unwavering commitment to preserving our planet's biodiversity. Thank you for being a part of this journey!

www.ingramcontent.com/pod-product-compliance
Lightning Source LLC
Chambersburg PA
CBHW070123230526
45472CB00004B/1386